Walk When The Moon is Full

Also by Frances Hamerstrom—

An Eagle to the Sky
(Iowa University Press)

Birds of Prey of Wisconsin
(Dept. of Natural Resources, Wisconsin)

Walk When The Moon is Full

by Frances Hamerstrom
illustrated by Robert Katona

The Crossing Press
Freedom, CA 95019

Library of Congress Cataloging in Publication Data

Hamerstrom, Frances, 1907-
Walk when the Moon is full.

(The Crossing Press series of children's stories)
SUMMARY: The author describes thirteen moonlight walks with her children and the nature observations they made.
1. Nature--Juvenile Literature. 2. Nocturnal animals--Juvenile literature. [1. Nature. 2. Nocturnal animals] I. Title.
QH48.H26 500.9'73 75-33878
ISBN 0-912278-69-2

Everything in this book really happened, but not during the course of one year, nor in one place. The actual locations were Waushara County, Wisconsin or Brush Hill in Milton, Massachusetts--January; Edwin S. George Reserve in Pinckney, Michigan–February, June and December; a composite of Fond du Lac, Wisconsin, El Paso, Texas, and New York City–the city chapter.

Each chapter was written on the appropriate night of the full moon. Most of the dialogue was made up; not, however,

The cory cat
with the cobby tail
walked straight into the salt blue sea
with a spider nesting on the end of its nose–
that's *you*!

Introduction

This story is about two real children, Alan and Elva, who called their father Hammy and their mother Fran. They lived on a 240 acre farm in Wisconsin and longed to go exploring at night.

Elva, the younger, liked to climb trees. Animals trusted her to come very close and her keen ears heard the faintest sounds.

Alan had a small but sensitive nose that found and remembered smells, just as his mind remembered facts. He liked to find things out.

Fran promised the children that she would explore with them twelve times–every month when the moon was full–for a whole year. Then Alan laughed, "The moon is full thirteen times a year." Hammy and Alan laughed because Fran didn't even know the moon is full every 28 days.

Each time they walked in the moonlight they found something new; sometimes they found animals eating, playing, or fighting.

Once Hammy took the whole family to the city and they were surprised at what they found there, too.

This book has thirteen surprises.

Fran Hamerstrom

ROYAL OAK

How it Began

Both children were in their pajamas for it was supposed to be the quiet time before they went to bed. Alan was looking at a book, but Elva had been standing at the window for a long time peering into the night.

Alan put down his book and went to the window too. He lingered there watching. At last he turned to his mother and sighed, "Do we have to go to bed early every single night until we are old?"

Fran set her mending aside and looked out the window. The moon was rising like a giant apricot, casting long shadows of gnarled oaks on the snow. Their mother was perfectly still for a long time. At last, she repeated Alan's words softly, "...every single night until we are old?"

Then she said, "No," in a faraway tone. Both children looked up at the sound of their mother's voice. "Why *should* children go to bed early every single night until they are old? I say, 'No!' "

"What do you mean, 'No'?" both children asked.

"I mean–no, you don't have to go to bed. The moon is full. Put on your sweaters and snow pants and overshoes. You can pull them right over your pajamas. Find your mittens. I think they are drying by the stove with mine. Just a moment, I'll tell your father we are going for a walk."

"A walk! A walk in the moonlight!" the children shouted.

January

"Sh! I hear something," whispered Alan. "Can you?"

They all listened. Moon shadows laced the snow. The cold of the night seemed to pull into the shadows–blacker and colder by the woods. Elva was on her hands and knees, moving her head near the ground. "I can hear you and Fran breathing," she answered.

"I heard something big, but it stopped. Elva isn't really *trying* to listen."

"I heard it," said Elva, "but I'm counting....sparkles. I can't pick them up. I have 51."

Alan moved away and crouched too. "I can turn sparkles on anywhere and nobody is counting mine."

Fran murmured, "No one in the whole world can say how many frost sparkles there are in America tonight–on grass, on leaves, and on the snow. Millions and millions and almost nobody knows they are there. But *we* do!"

"Don't grownups know they are there?"

"No, Darlings, most of the grownups are indoors and most of the children are in bed. Shall we take one little leaf and count the sparkles?"

"No," Elva spoke unexpectedly. "Listen! I hear something again. There are big animals–moving."

A branch snapped beyond the marsh, and then twigs swooshed.

Elva said, "Big animals *are* moving. I can hear their feet on the ground."

Every now and then, they heard the willows shift and snap back.

"Deer," said Alan. "They're pulling at the bushes to browse."

"Shall we go nearer?" Fran whispered.

She took Elva's hand and helped her over a log. Elva's snowsuit squeaked as Fran pulled her over the scratchy bark.

"Too much noise," Alan muttered.

They all stood still. Alan was right. The thunder of hoofbeats pummeled on the frozen ground as the deer tore down through the marsh and into the woods.

"Can we get near them?" Elva asked.

Fran said, "Well, we could put some apples on the window sill tonight and see if deer come to eat them."

"Let's go home." Both children wanted to hurry back.

When they got to the house, Alan looked at the big basket of apples on the kitchen floor. "How many can we take?"

"Six apiece," Fran answered.

The children ran outside and brushed the snow off the living room window sill and set twelve apples carefully in place.

"Will the deer come to eat them?"

They did!

Long after the children were asleep, Hammy heard the deer by the window. Fran put down her book and tiptoed into the children's room; "Wake up," she whispered. "Follow me."

Hammy stood like a statue, staring out of the window. Then the children saw them too.

There were two deer just outside the window—a doe and a big yearling. The yearling reached gingerly toward the window sill. His nostrils loomed huge as he opened his mouth, turned his head sidewise and took an apple. For a moment his breath formed a misty cloud on the window pane.

Then the doe reached for the last apple and chewed it. At last both deer turned and swung away into the night at a slow trot. Just before they disappeared, the yearling paused and looked back at the house. A young deer was memorizing a place where he could find a delicacy—apples.

February

"Oh Alan, I can walk on the top of the frozen snow and the moon is full. Call Fran!"

"Fran, the moon is full, come!"

"I'm working, Darlings–let's go tomorrow night."

"Oh, please come; it might rain again."

Fran got up to look at the thermometer–10 below! The sparkle of the ice storm frosted the trees and moonlight struck the barn roof white. "You are right, Darlings, I don't think it will rain tomorrow but tonight is the night to walk. We'll walk on the crust of ice."

Alan called, "I can slide on the crust! Let's go to the hill and we can sit down and coast."

At the top of the hill Fran and Alan sat down and started sliding carefully, but Elva threw herself on her stomach and gave a big push. Down, down, down they went, faster and faster. Elva beat them to the bottom of the hill, but suddenly she wailed, "I've lost a mitten."

Fran said, "Wear mine, and I'll go back and look for it. You two stay here and wait."

The children could hear her slowly, slowly, crunching up the slippery hill, till the sounds became fainter and fainter; then all was still. They were alone.

Moonlight shone on the glare-ice crust. The birches, heavy with their load of ice, were bent so low that some whole trees arched over. Their tops drooped down to the ground and were trapped by crusted snow.

"Alan," whispered Elva, "There's an animal."

"Sh! I see it too, down under the birch tree top."

Just then there was an enormous noise-Fran was sliding down the hill bringing Elva's mitten.

"Oh," said Elva, "You scared it. There was an animal, it came--"

"Where?"

"Under the birch top–it's gone–it ran away."

"Let's look," said Fran.

Under the tops of the bent-over birches lay rabbit droppings like tiny marbles. The rabbits had eaten the twigs at the top of the birch trees!

Elva asked, "Will the birch trees always stay bent over?"

"No," Fran said, "When the ice melts they'll straighten up again and no one will know the difference next summer."

"I will," said Alan. "Shall we keep it a secret? I'll ask Hammy what has eaten the tops of the birches."

"Your father is a woodsman and hard to fool, but let's see what some-one from the city says when we tell him rabbits have eaten our treetops."

When Grandma came to visit in spring, the children took her straight to the birch trees and Alan asked, "What ate those treetops?"

Grandma looked up at the ragged tops of the birch trees and said, "Squirrels of course."

"Fooled you! Fooled you!" both children shouted. "It was *rabbits* and we watched them."

March

"I choose! We go to the pond!" shouted Alan.

"Not tonight, Alan. The ice is too thin to hold us–and don't shout so."

"I choose–and I *want* to go to the pond!"

"No, Alan."

Alan was furious and forgot to be respectful to his mother. He yelled,

The cory cat
with the cobby tail
walked straight into the salt blue sea
with a spider nesting on the end of its nose–that's *you!*

"Alan."

For a few moments, it looked as though nobody was going for a walk in the moonlight. Elva sat on the floor with one boot on and the other in her hand, waiting to see what was going to happen next.

Hammy stepped out of his study and said, "There were possum tracks in the long fence row today."

Nobody said anything more. They just finished pulling on their boots, put on the rest of their clothes and headed for the long fence row.

It was the darkest full moon night of winter for much of the snow had melted. Wind roared above them in the treetops, but down near the ground it seemed quiet, almost as though they were in a cave.

Elva found a hole in the earth–a hole with frosty edges. "There must be an animal inside. Do you think that's where the possum went?"

Fran called Alan who had the best nose in the family. They all kneeled on the frozen ground and sniffed the hole. Alan raised his head immediately, "I'm surprised you can't tell. There's a fox down there. It smells just like the fox Hammy brought home last winter."

Elva spoke softly, "Is it asleep?"

"It probably is." Fran continued, "We can't track anything tonight. And we can't even *hear* anything with all this wind."

"*I* hear something." Elva was looking straight up. Two long icicles hanging from branches were tapping against each other above their heads. Fran reached up and broke one off for each child.

Elva sucked hers, bit off its tip and held it in her mouth. "It's sweet," she mumbled, "Sweet on the bottom."

Then she climbed quickly, grasping the swaying branches, to get another.

Alan stood still. Slowly, the tip of his tongue explored the strange taste of the chilled surface.

"Why are the icicles sweet?" he asked.

Fran thought a moment. "That old maple's branches must have twisted and cracked in the winter storms–now the sap is leaking out and freezing."

"Maple? Can we make maple syrup?"

"Yes." Fran struggled out of her big parka, spread it out and started pitching icicle tips into it. Alan and Elva added more. Shifting icicles caught sparkling surfaces on the moonlit heap. At last it was so heavy that Fran could hardly carry it home.

Fran boiled the icicles in a bucket. It took a lot of boiling, but at last everybody had pancakes with real maple syrup.

April

"Shall we go for a walk? I heard the sound of a woodcock peenting–its mating call."

The children looked at their mother as though she were absolutely silly.

"It's not dark yet,"they protested.

"I know," she answered, looking intently out of the window at the fading day. "I think we'd better get started right now."

Both children objected: "It's too early to go."

"Get your raincoats, Dears."

"Raincoats?" Alan protested, "It's not raining!"

"Hush down. Now let me explain. We are going to try to sneak up on a woodcock and watch it peenting on the ground. You don't have to *wear* the raincoats, they are just to hide under. If we don't start now, we may not find a woodcock. It's almost dark."

They headed straight for the abandoned field, where dark pines and white birches grew. They lifted their feet high where the grass was deep and only slightly flattened by the winter's snow. When they got to the clearing, where mosses and low sedges covered the ground, they stood still to listen for the "Peent--Peent" of a woodcock.

Fran said, "Listen carefully. I'll squeeze your hands when it is time to talk."

Elva's hand kept squeezing and tugging, till Fran gave the signal. Elva spoke first. "I hear five or maybe six. One by the house and one by the barn and one by the machine shed and more toward the pond. And one right here."

"I hear *one*," said Alan, "Behind those two pines."

"If it's any comfort to you, Alan, *I* wasn't sure I *heard* any, but I *saw* one go down behind those pines. Who could hear the twittering flight song when the woodcock went up?" Fran asked.

"I could," said Elva.

"Elva is the youngest member of the family," said Fran, "But tonight she leads the expedition. We need her good hearing. When his peenting stops and the flight song starts, we'll run almost to those pines and hide under our raincoats. When Elva says the flight song has started, we'll rush to where he peented and hide again."

Brambles caught at their clothes; they scuttled when the woodcock's flight song started. But in the blackness of the night they scurried too far. The woodcock–dark, small and swift–swirled down out of the sky in a great sweeping arc right *behind* them. "Peent–Peent."

He strutted on the ground by a big old ant hill. And then, swift-winged, faster and faster in upward curves, he flew toward the starry sky again.

"Let me have one of your shoes, Elva! Quickly!"

"Why?"

Fran didn't answer. Elva handed her a shoe and Fran ran crouching toward the ant hill, leaving the shoe near it. Not until she was back hiding with the children did she whisper, "Maybe he'll think it's another woodcock."

And he did.

He swooshed over their heads and lit near Elva's shoe. He raised his wings right over his head and dashed at it in little runs.

"He touched it, Fran, he touched my shoe!"

Swifter than a swallow, the woodcock swooshed past them and disappeared into the black night.

It was time to go home and getting colder. They were glad they were wearing rain gear for warmth. They pushed past brambles and through sweet fern toward the path. Gradually, as they walked, the night got lighter until the faint orange glow of the rising moon guided their footsteps home.

May

"Where are we going tonight?"

"Well, it's rather cloudy and Alan had a splinter in his foot."

"Please, we want to go—we could stay on the road."

Alan said, "I can hear the cars on the highway—let's go there."

"To the highway, Alan? What a funny idea!"

But they did go toward the highway, past the marsh with the singing frogs. Suddenly it started, "Twash—Twash"—a harsh insistent call such as no small animal could make.

"Do you still want to go to the highway or shall we sneak up to that sound in the woods?"

"Twash--Twash."

"Sneak it!" both children whispered.

Slowly they walked deep into the dark woods. A car passed on the highway. The "Twash" stopped and the moon hid behind a small cottony cloud.

"Look!" Elva said, "A big bird!"

Then moonlight caught the fluffy down on a big owlet. His new, grownup feathers gleamed dark, peeking from his downy whitish body. His eyes, big in his big round head, blinked and blinked again—slowly.

"Let's sit here and watch him. We can't stay long or his parents won't come to feed him. They will be afraid of us." Fran explained, "You see, he's calling for his supper."

"Twash–Twash."

Then the baby owl stopped calling. It was looking toward the west. Soon they could hear the faint sound of a car coming. The headlights grew brighter and brighter. The owlet turned its head slowly as the car passed–watching it–and then, long after even sharp-eared Elva could no longer hear the car, the big downy owlet looked east.

"Twash–Twash." They sat and watched. Again the owlet stopped calling.

"Fran," said Elva, "There's another car coming. The baby has turned its head!"

They listened–first there was silence–then a faint hum and at last the car roared past.

"Someday, when you both have cars and are driving at night, you can remember as you pass the woods in May, that many baby horned owls are turning their heads as *you* go by."

June

The children had been in bed for two hours, but Alan burst out of his bed to ask, "What's going to happen tonight?"

"Alan, I never know what's going to happen. Right now the clouds are over the moon. Do you still want to go walking?"

"You *promised* us," wailed Elva, "The moon is full even if we can't see it very well. You know it's full."

Fran said, "All right–I promised you–we'll go walking."

"Barefoot?" asked Alan.

"All right, barefoot, but I'm going to take Elva's shoes along."

Their feet moved through the dew laden grass. Barefoot, they moved through the darkness more slowly than they would have with shoes on. As they climbed down the steep slope into the marsh, fireflies flashed below them. Fran put Elva's shoes on. It was getting cold.

Fran lay down in the damp grass to watch something she had never seen before: the flash signals of thousands of fireflies in an *even layer* above the water of the marsh. The fireflies had found just the right temperature for signalling each other!

When they reached the marsh, the children ran ahead. Blinking lights flickered above them. Fran held Elva up so the carpet of fireflies lay below her and then lowered her so fireflies hovered above her like a ceiling of lights.

Alan and Elva slipped away in the darkness of the windless night. They were making unusually little noise.

When they came back, wading in the shallow water, Fran found herself shivering.

"It's time to go home."

To her surprise both children said, "Yes."

As they clambered up the steep trail homeward, Fran offered to hold Elva's hand.

"No." Elva pulled her hand away.

Both children were giggling. They plainly had a secret.

Fran was puzzled until she and Hammy went to bed. When she turned the lights off, she found out what they'd been up to. The children had let fireflies go in the bedroom!

Fireflies paraded over the bedside table blinking their lights.

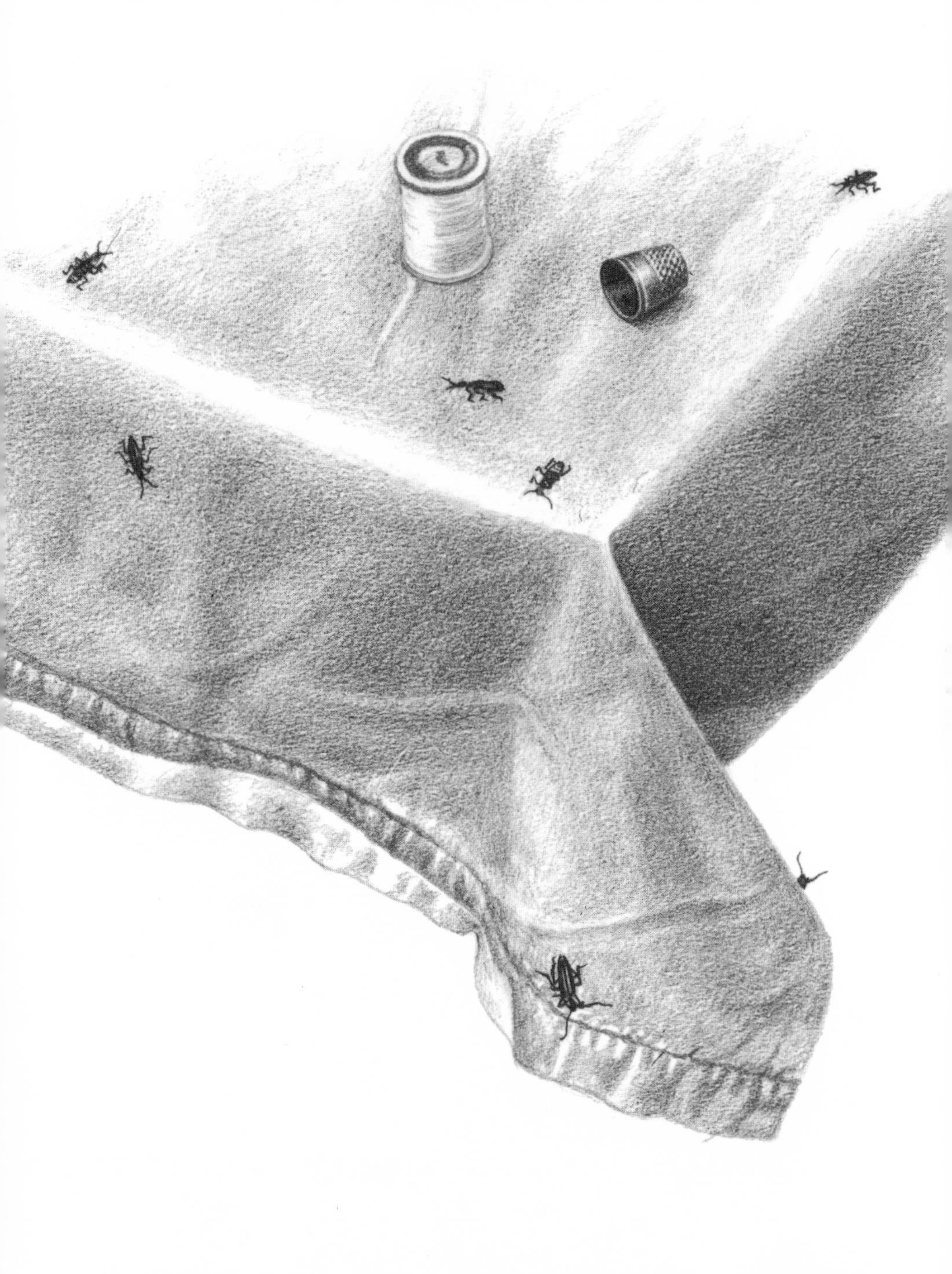

July

The children came into the room looking very pleased with themselves.

"You look as though you knew where you wanted to go tonight," Fran smiled.

"We don't care where we go. We want to take flashlights!"

"But Darlings, won't it spoil everything? Just think, it won't be the same."

"*Why* can't we take flashlights? Elva can use mine and I'll carry the big one."

"Well," Fran agreed reluctantly, "We'll take them along, but let's walk as far as the pond by moonlight because I'd like to."

(She had no idea that *she* would want to use a flashlight before long.)

Alan ran ahead to the pond and soon he was shouting, "I've got my flashlight on."

Elva stooped in the shadow of the trees to smell the sweet spicy fragrance of white cockle flowers–wide open at night.

"We can see fish–they never go to bed," Alan called.

Elva added, "And flowers don't either!"

Fran said, "Some flowers do."

"Show me a flower asleep," quipped Alan.

They went to the weedy edge of the cornfield where the morning glories grew. All the flowers were twisted tightly shut.

"Let's see if they'll open by flashlight!"

The children held their lights right next to a morning glory but it wouldn't open.

Then they went to the roadside. But the spiderworts wouldn't open either. The daisies were open wide.

Suddenly Elva exclaimed, "The clover is gone!"

Fran *knew* that there was a clover patch right by the spiderworts. She borrowed Elva's flashlight, but she could find no clover.

Alan got on his hands and knees and looked carefully. Then he cried, "Here it is! There's lots of it. Did you know that clover plants fold their *leaves*?"

"No," Fran answered, "I didn't, and I think your father will be surprised to learn this."

"Bring him some! Let's bring him some!"

"Alan, can you find your way back to the morning glories?"

"Of course I can."

"Then you pick a morning glory and Elva and I will pick a spiderwort and some clover leaves. We'll put them all in a vase and see what they look like at breakfast."

Elva picked lots of clover leaves.

In the morning, Alan whispered to Fran, "Will Hammy look at clover *leaves*?"

He did.

At breakfast, the morning glory was curled shut, the spiderwort's blue flowers were partly awake, but the clover leaves had opened wide.

All three waited in the doorway to see what Hammy would say.

He examined the strange bouquet. Suddenly he looked up and smiled. "Where did you get the four-leaf clover?"

PEACE

A Walk in the City

"We are going to the city for a few days."

City? Alan and Elva were speechless with surprise. They had almost never been in a city.

"Where will we sleep?" asked Elva.

"In a hotel."

Finally Alan exclaimed, "But the moon will be full. Will we walk in the moonlight?"

Fran answered uncertainly, "We'll try."

The very next night Fran and Alan and Elva left the hotel room and went down to the lobby in the elevator. Then they passed through revolving doors to the street.

Cars and busses whizzed by. Bright lights flashed from neon signs and men at curbs called, "Taxi! Taxi!"

These children—at home in the country—dared not let go Fran's hands as she pushed her way along the crowded sidewalk.

The sky was a haze of colored moving lights. Fran led the way to a quieter street where it was darker. They stopped and looked up to try to find the moon.

A man came out of a doorway and looked up too. "What's going on?" he growled.

Elva answered politely, "We're all looking for the moon."

The man just laughed and went back to the doorway.

Alan called after him, "If we could get up high, we could see the moon."

A few minutes later Fran noticed an old man locking a church door. "Run!" she shouted.

They ran to the old man and breathlessly, Fran asked him if they could go into the church and up into the steeple.

He was so surprised that he hardly knew what to say. Both children said, "*Please*!!"

Then the old man chuckled, unlocked the huge door and led them between the pews to the rear of the big, old church.

"Bell tower's up there Ma'am," he murmured respectfully. "My legs are tired. Take my flashlight. I'll wait here."

First they climbed up a carpeted stairway.

They came to narrow, dusty wooden stairs that went round and round. Gradually, the sounds of the city faded away.

At last they climbed into a little room with high, narrow windows. Alan took charge of the flashlight and Fran held Elva up high so she could look. Elva couldn't see the moon anywhere–just hazy moonlight and high up a maybe moon.

Suddenly Alan whispered, "I've found a door." He twisted the rusty knob and pulled the door open.

Cool night breeze blew into the little room from the base of the bell tower. They stepped outdoors. Arches supported another room high above them.

An iron ladder led up through a hole in the floor of the room above.

Elva started up the ladder, but Fran said, "Wait, I want to make sure it's safe."

Now Fran took the flashlight and climbed part way. "The ladder is fine, but our city shoes are too slippery." They all took off their shoes and climbed barefoot.

Elva climbed first, then Alan–and Fran brought up the rear. They had often climbed ladders picking fruit, but not at night, high over a city.

When they were almost at the top, a big, almost white owl flew down through the hole in the floor and out into the night.

Fran said, "Stay where you are. I'm going to explore." She climbed past the children and disappeared through the dark hole. They saw the beam of her flashlight in the dusty air in the room above.

"Come," she called, "Look what we've found!"

There was a sharp hissing from one corner of the bell room. The children would have been frightened, except that Fran sounded so happy.

Eight fluffy baby barn owls hissed, threatened and tipped their heart-shaped faces at the intruders.

But the smallest owlet was too little to be left without its parent for long, so Fran and the children climbed down quickly, thanked the old man who had let them in, and walked back to the hotel.

When they got back to their father, he looked at the happy dirty faces and said, "You have stars in your eyes and cobwebs in your hair. Did you find the moon?"

August

"Fran, I think the children are up–whispering outside our door."

"Up? At this hour? I'll talk to them."

Fran opened the bedroom door and both children raised expectant faces.

"You forgot the *moon*," Alan whispered. "It's full."

"You mean you want to go walking *now*? It's midnight!"

Both children nodded their heads.

"All right,"Fran answered. Let's go quietly to the pond. It's too hot to sleep anyway."

Elva exclaimed, "I hear four frogs going 'chunk-chunk'!"

"Let's get in the water and sneak up on a frog," said Alan.

So they all slipped off their clothes and walked softly into the warm water and lay down so only their heads were sticking out. Using their hands, they pulled themselves through the bullrushes toward a chunking frog. As they sneaked closer, great arcs of silvery light spread far into the pond.

They crawled nearer and nearer, barely disturbing the water, until the frog boomed so loud that it hardly seemed possible it was just a frog. All three faces were within two feet of him when–*plop*!–he was gone. One shiny bubble floated where the frog's head had been. The bubble drifted slowly toward Elva's face and burst.

"I want to catch a frog," said Alan, but Elva wanted to leave.

"Do you want to come home *alone*?" Fran asked.

"Sure," said Alan.

Alan sneaked one frog after another but they all jumped long before he could grab any. He waded ashore. After he pulled on his clothes, he saw a pinpoint of light in an upstairs window. Fran must be putting Elva to bed.

He started walking home. The path of the silvery moon on the pond pointed straight toward him. He walked a little way and it still pointed toward him. He walked back and forth along the shore and wherever he went, the moonpath pointed at one person—Alan.

Later when Fran kissed him good night, Alan whispered, "I found something out when I was alone. The path of the moon doesn't always stay in the same place. Wherever I went, it followed *me*."

September

The house was quiet. Hammy was rummaging softly in his study and everyone knew that he was not supposed to be disturbed when he was working. It was past bedtime, but the moon was full.

Alan said, "You look for Fran all over downstairs, and I'll look upstairs. The car is here, but I can't find her."

The children went into each room of the big, old farmhouse but they could not find her anywhere. At last they interrupted their father politely and asked, "Where's Fran?"

He answered promptly, "She's picking flowers."

Picking flowers at night? The children didn't have permission to go wandering around outdoors at night alone; they went outside rather quietly, and sure enough, Fran was in the garden picking flowers as quickly as she could.

Elva asked, "Why do you pick flowers at night?"

"It is 34 degrees right now. Frost will come tonight. Most of the flowers will be dead by morning, except for those that we put in vases. I've just finished picking–don't worry, I know the moon is full."

"We want to swim in the pond again, just like last month."

"Put on your jackets. We'll have a look at the pond."

The grass was very wet. Elva asked, "Will we be wearing mittens soon?"

Alan repeated, "We want to go into the water just like last time."

Fran said, "Let's see." She took Elva's hand and all three walked through the deep shadows of the trees, across the open meadow and straight to the pond.

"Off with your jackets and things!" Fran sounded cheery. "I'll help Elva. Quick! Everybody into the water."

The air was cold.

But the water was warm! Mist hung over the pond and moonlight tipped the top of the mist. Before they had time to object, both children were in the pond. When they kicked their feet, big bubbles rose. No frogs called, but the water was so warm—almost like August. Both children had been sure that Fran would refuse to let them go into the water, and here they all were, lying under the mist in the moonlight.... kicking up bubbles that drifted like domes across the smooth surface of the pond.

"Are you warm enough, Darlings?"

Both children laughed.

At last it was time to get out of the water and go home. The air was stinging. Fran helped both children bundle up quickly and all three ran back to the house, breathing sharp whiffs of winter as they ran.

The next morning almost all the flowers in the garden were dead.

October

"Let's follow the fence row by the cornfield."

Both children put on their winter jackets, but Fran forgot hers. Though the wind blew through the tops of the tall pines as they started out, it was not cold deep in the sheltering woods.

But when they crossed the road to reach the fence row, the wind whipped across the open fields and made their faces tingle. Fran said, "If we're going all the way to the fence row, I'll run home for my jacket. You walk down the road and I'll catch up before you get there."

Alan and Elva heard Fran running, and then all was quiet. They were alone except for the moan of the wind.

They wandered slowly down the dark road.

Alan said, "I smell an animal."

Elva moved closer to Alan and whispered, "I see him."

She pointed at a shape not far away. They walked a few steps closer, but the animal didn't run.

The animal did not move.

Alan and Elva stood perfectly still. White clouds passed over the face of the moon and the wind roared in the treetops. They listened for Fran, and before long they heard her running toward them.

Alan hushed her, "Sh! There's an animal on the road."

"Show me."

They tiptoed over to the animal. The moon shone on a dead possum. Fran said, "This is pretty far north for possums; maybe the natural history museum will want to have it."

"How old is it?" asked Elva.

Fran held it up by the tail.

"I can't tell by moonlight. Shall we take it home and give it to the museum?"

"I want to carry it first," said Alan.

After a while Elva carried it.

"We're helping a museum," she said proudly.

At home they laid it on the kitchen floor. Fran looked intently. "It's an old possum—more than a year old. Look, the tips of its ears and the tip of its tail are gone—frozen off. Possums that spend the winter here always lose part of their rounded ears and long pointed tails."

"Can we take it to the museum now?" Elva asked.

Alan answered, "Museums are closed at night, even when the moon is full."

Fran promised, "We'll take it in the morning."

And they did.

The man at the museum said, "This is the first record for the county. We are glad to get this specimen. Who shall I say donated it?"

"Alan and Elva Hamerstrom."

The museum man said, "Thank you very much."

MUSEUM

November

Just as Fran and Alan and Elva started across the soft new snow to walk in the moonlight, a car drove up the road and turned in the driveway.

"Company," both children groaned. "Does it mean we can't go?"

"Wait here," said Fran. "I'll tell them I have to be gone for a little while."

The children heard the sounds of grown-ups greeting each other at the door. Then all was still again.

Suddenly Elva asked, "Do you hear the dogs?"

Alan couldn't hear anything except the windmill creaking now and then. It seemed to the children that they waited a long time and that maybe Fran wasn't coming. They stood very still. After a while Alan said, "I hear dogs barking–I can hear them."

Then Fran bounded out of the house saying, "They really wanted to see your father."

"Listen," said Alan, "Dogs."

Fran looked at Alan in surprise because he seldom could hear distant sounds.

"Elva heard them first," he admitted, "But I can hear them too. Dogs, can *you* hear them?"

Fran listened carefully. The sound of high excited yelps drifted toward them. Then she nodded her head. "Yes, I can hear them, but I'm not so sure they are dogs. Maybe they are foxes."

Not another word was needed. They all walked quietly toward the sound, which seemed to come from the pond.

The first heavy snowfall of the season lay in powdery drifts and the frozen pond gleamed white.

Clouds covered the moon and then parted. Moonlight shone bright on the pond. Two animals with bushy tails tugged and pulled at something on the ice, romping like puppies. Then they barked.

"Foxes," Fran whispered, "I'm quite sure they're foxes. And they've got something out there–maybe an old, dead muskrat."

"Why?" Elva asked.

Alan answered promptly, "It's their supper. Foxes eat meat."

The clouds drifted over the face of the moon again, almost hiding the dark shapes of the two animals.

"Are you sure they're foxes?" Alan asked.

Fran thought for a moment. "I'm *almost* sure. Tomorrow, after breakfast, we'll go look for their tracks."

And they did.

A crow was picking at something way out on the thin ice where the animals had been the night before and the surface of the pond was laced with fox tracks.

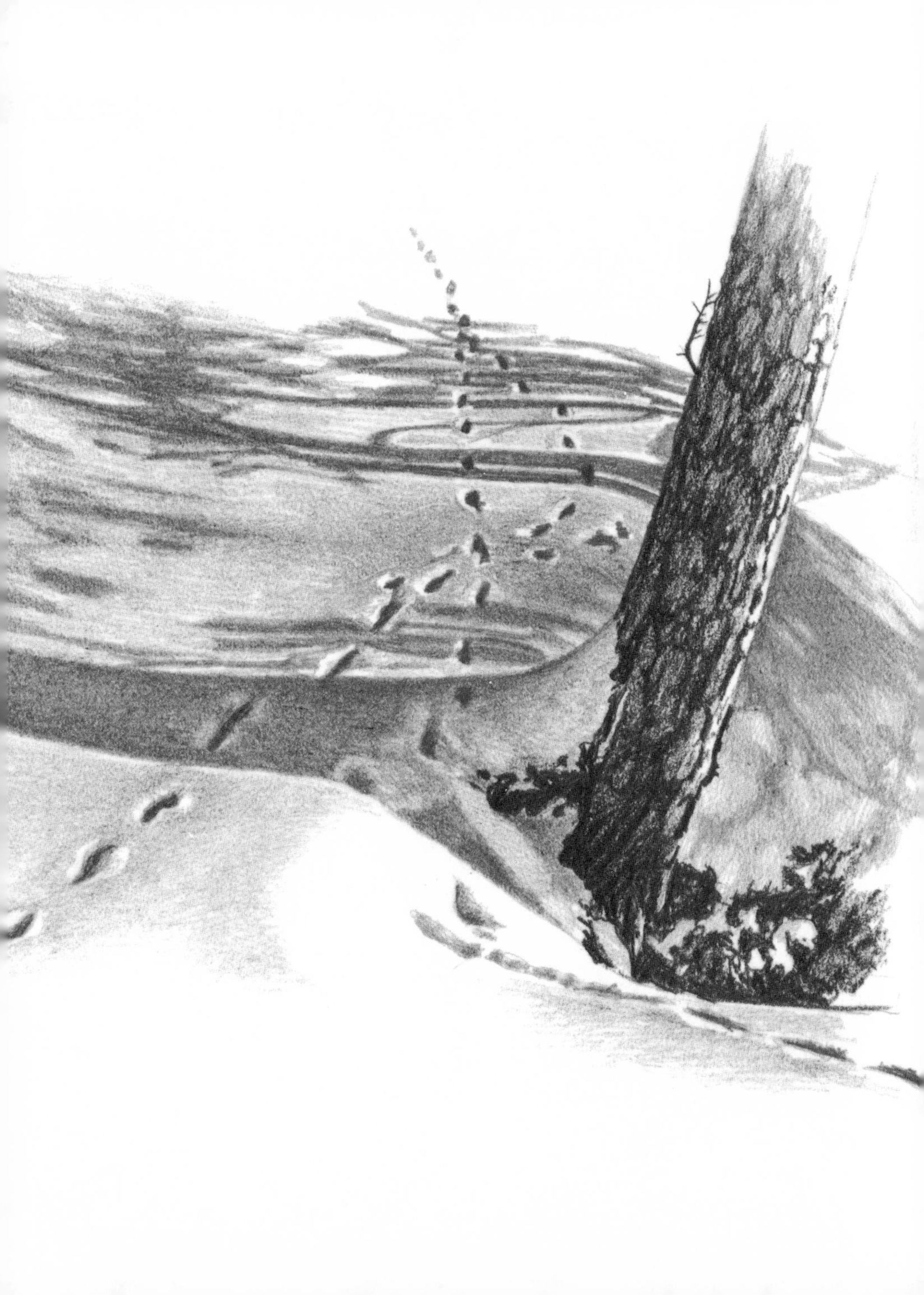

December

The grown-ups were together in the study talking in low tones. The children listened. Fran was saying, "Well, we could stay indoors. We could turn out all the lights and look out of every window."

Alan stopped tying his bootlace and listened more carefully.

"...or we could toast marshmallows."

Both children knew that there just might not be a walk in the moonlight, even though Fran already had her outdoor clothes on.

Hammy went over to the window and peered at the thermometer.

"Thirty below," he shook his head.

Alan spoke up immediately, "Comfortable cruising weather."

The grown-ups burst out laughing. Alan was quoting from his favorite Eskimo book.

Fran pleaded with Hammy, "We won't go far–just to the Little Marsh and back." He finally agreed.

Then they put on their warmest clothes and took the trail toward the Little Marsh. Bare trees cast clear cut shadows across the path, and Alan and Elva jumped over the shadows as they bounced along the trail.

Then they stood on the hillside overlooking the marsh. Thirty below is cold and strange. They stood still a long time.

Suddenly Elva gasped, "Sh! Look at them! Look down there!" Small shadows were romping and playing all over the white surface of the marsh.

"Weasels," whispered Fran. "Weasels are white in winter so we can only see their shadows. If we could get close to one, we could see its dark eyes."

They sneaked down to the marsh, making as little noise as possible, but in a twinkling all the weasels disappeared. One ran down a hole inside a tamarack root right near them.

Elva said, "Listen, somebody is coming!" And sure enough, somebody was crunching through the woods.

It was Hammy!

He was glad to find them and said, "You've been gone a long time for such a cold night. Do you realize it is 32 degrees below zero right now?"

Everybody told Hammy about the weasels, and then Hammy asked, "Do you know where the weasel went down?"

Alan and Elva took him to the tamarack root, and Hammy motioned everyone to crouch. Then he squeaked like a mouse.

The weasel popped his head right out of the hole. His little leathery nose gleamed in the moonlight and his whiskers quivered.

Elva giggled with delight.

And he was gone!

Epilogue

Walk when the moon is full.

My children, Alan and Elva, will never forget exploring in the moonlight.

Years later, Alan admitted to Elva that he was sometimes scared when they went moon walking and Elva answered, "I was sometimes scared too—especially when Fran left us alone. But I know that some day I will walk with my children when the moon is full."

Elva's daughters, Becky and Lita, to whom this book is dedicated, are now just old enough to capture the magic that is for anyone--a walk in the moonlight.